NEPTUNE

Other Voyage into Space Books

A Voyage into Space Book

NEPTUNE

Voyager's Final Target

by Franklyn M. Branley

HarperCollins*Publishers*

My thanks to so many people in the preparation of this book. Special thanks to Dr. J. Pieter deVries of the Jet Propulsion Laboratory for his critique and valuable suggestions.

—F.M.B.

Photo Credits:

Smithsonian Institution Photos: No. 56,809 (p. 14), and No. 57,433 (p. 17); Yerkes Observatory: (p. 16).

All other photos and graphics courtesy of NASA/JPL

Library of Congress Cataloging-in-Publication Data
Branley, Franklyn Mansfield, date
    Neptune : Voyager's final target / by Franklyn M. Branley.
        p.   cm. — (A voyage into space book)
    Includes bibliographical references and index.
    Summary: Details the activities of the American Voyager 2 space probe as it made its 1989 flyby of Neptune and its moons. Discusses the eighth planet's orbit, atmosphere, rings, and geology.
    ISBN 0-06-022519-X. — ISBN 0-06-022520-3 (lib. bdg.)
    1. Neptune (Planet)—Exploration—Juvenile literature.   2. Voyager Project—Juvenile literature.   [1. Neptune (Planet)—Exploration. 2. Voyager Project.]   I. Title.   II. Series.
QB691.B73   1992   91-2469
523.4'81—dc20   CIP
                                                                            AC

Contents

1. The Final Mission of Voyager 2

On August 24, 1989, after a twelve-year journey of more than four billion miles, the space probe Voyager 2 made a close approach to the planet Neptune. At the Jet Propulsion Laboratory (JPL) in Pasadena, California, the Voyager team watched their video monitors intently as the first pictures of Neptune, taken by Voyager's special TV cameras, came in. Since the probe's launch in 1977, the team had worked together closely. Together they had solved complex technical problems; they had cheered Voyager's success at Jupiter, Saturn, and Uranus; and now they were seeing Voyager's final triumph. They had become almost a family. Some team members had retired, others had gotten married

This is a computer-simulated view of what Voyager 2 must have looked like as it passed Neptune and one of its satellites, Triton.

and had children, still others had become professors at colleges and universities. In their personal lives, many changes had occurred, yet the group was still a team, and they had waited a long, long time for this exciting moment. Certainly it was one of the greatest achievements of the space age. Since Neptune's discovery in 1846, virtually nothing had been known about this mysterious planet. Voyager 2 has given us enough information to fill many books and to keep scientists busy for years to come.

When Voyager 2 was launched, few expected that it would ever reach Neptune or, even if it did, that it would still be able to take pictures, gather information about the planet, and send all the information and pictures to Earth. The probe was designed to operate for four or five years, to survey Jupiter and Saturn, which it did.

But it did a lot more. If all went well, scientists knew it would be possible also to visit Uranus and Neptune. The team worked hard to keep Voyager going. After it had done its job at Jupiter and Saturn, Voyager was still working so well that the scientists decided they could expand its mission. They would change its path slightly and move it into an orbit that would carry it farther than any space probe has ever gone. Much to the pleasure of the Voyager team, the probe's instruments were still working when it reached Neptune— eight years after it had left Saturn.

Neptune is the eighth planet in the solar system. Pluto, the ninth planet, varies greatly in its distance from the Sun. Usually it is the most distant planet. At the point in its orbit where it is nearest to the Sun, however, it is closer to the

Sun than Neptune. At the time of Voyager 2's visit, Neptune was actually the outermost planet. (Pluto was too far away from Voyager's planned path to be explored.)

Voyager 2 sped toward Neptune at some 40,000 miles an hour. It traveled in a wide, curved path 4.4 billion miles long. If it had gone straight toward Neptune, the journey would have been 2.7 billion miles. Space probes move in curved paths, however, because for the most part they are coasting, not in powered flight. For powered flight a tremendous amount of fuel would be needed—more than the probe could carry. When Voyager was at Neptune, it was so far away from Earth that radio messages took four hours to get here, even though radio signals travel at the speed of light (186,000 miles a second).

Photographs taken by Voyager's cameras were changed to radio signals and sent to Earth much the same way TV pictures are changed to signals that are broadcast to your receiver. Voyager's radio transmitters operated on 22 watts of power, which is equal to that used in a dim light bulb. By the time the signals arrived at Earth, their strength had dropped to less than a billionth of a watt. Large antennas strategically placed around the world picked up these weak

Large antennas, located in California, Spain, and Australia, picked up Voyager 2's weak radio signals. The signals had traveled more than 2.5 billion miles, from Neptune to Earth.

Voyager's radio signals were transmitted from the antennas to the JPL's Space Flight Operations Facility (SFOF), where they were turned into pictures. Instructions to Voyager were sent from the command center at the SFOF to the antennas, which then beamed the instructions to Voyager.

signals and sent them to JPL. They were made stronger and fed into computers that changed them back into pictures.

Scores of space probes have been launched over the years. The United States, the Soviet Union, Japan, Europe, Israel, and India have each developed probes. Of them all,

6

Voyager 2 has given us the most spectacular views of the outer planets. Voyager 2 and its twin, Voyager 1, were not the first space probes. They grew out of the less powerful Mariner probes that had been sent to Venus and Mars.

Information gathered by a probe is not sent to Earth at once. It is stored in a computer. After a certain amount of information has been collected, and when the probe is in the best position for transmission, the information is sent out. In Mariners and other earlier probes the computers were slow, taking eight to ten hours to collect and transmit data. Improvements in computers reduced collection and transmission time to less than an hour. Now, mini super-computers aboard probes can send all their information in only a few minutes. We get more information, and we get it faster than ever before. Compared to probes launched in the 1980s and '90s, Voyager took quite a while for a transmission. However, no probe has traveled as far as Voyager 2, and none has given us so much new information about our solar system.

No matter what the target may be, it takes a lot of planning and good engineering to design a probe, build it, and get it to its target. If the probe's destination is the outer planets—

Jupiter, Saturn, Uranus, and Neptune—the planning and engineering become even more demanding. For one thing, aiming the probe and keeping it on course become extremely difficult. The slightest error in direction, continued and amplified over billions of miles, amounts to a huge error in final destination. Also, remember the probe is going 40,000 miles an hour. Even if an error in direction is corrected very quickly, the high speed of the probe will have carried it thousands of miles off course.

Another problem is power—providing the electricity to run the computers, cameras, radios, and other equipment. Probes that remain close to Earth and the Sun can use solar power. They are equipped with cells that change sunlight into electricity. Probes going into the outer solar system, however, cannot use solar cells. As a probe travels farther from the Sun, the amount of sunlight that reaches it decreases. Voyager 2 used electric generators driven by the heat from nuclear fission. The fuel in the generators was plutonium. In 1989 the generators had been operating for twelve years, and judging by their performance at that time, engineers predicted they would furnish enough electricity to run a useful mission for another twenty-five years.

During the course of its long journey, Voyager 2 had several technical problems that put its mission in danger. But even though the craft was millions of miles from Earth, engineers were able to solve those problems and keep Voyager 2 going. In 1986 Voyager 2 arrived within 15 miles of the point, 17,980 miles from Uranus, that the JPL navigators had selected. And it was on time within one second. When Voyager 2 arrived at Neptune three and a half years later and 1.5 billion miles farther away, it was still right on target—within 3,048 miles of Neptune's north pole. And it was right on time.

Voyager 2's twin, Voyager 1, was sent to Jupiter and Saturn. While at Saturn, the path of Voyager 1 was changed slightly so it could get a better look at Titan, the largest of Saturn's satellites. That change made it impossible for the craft to visit Uranus and Neptune. Instead, Voyager 1 left the solar system and traveled into interstellar space. Earlier, the probe had discovered eight active volcanoes on Io, one of Jupiter's satellites. These were the first active volcanoes to be seen anywhere in the solar system outside of those on our own planet. Later Voyager 2 discovered a totally different kind of volcano on one of Neptune's satellites. Be-

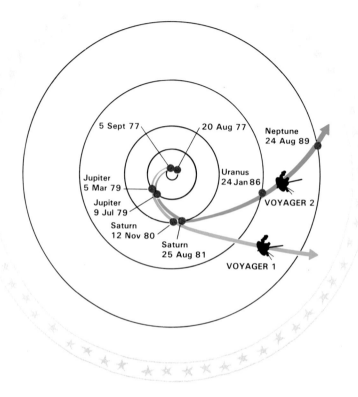

Voyager 2 completed a 4.4-billion-mile journey to Neptune and now is moving out of the solar system. Its twin, Voyager 1, left the solar system after visiting Saturn.

cause of the Voyagers, we know that the four largest planets—Jupiter, Saturn, Uranus, and Neptune—have rings. Also, because of the Voyagers, we know there are a lot more satellites in the solar system than we had previously believed. There appear to be sixty-one altogether—though there may be others still to be discovered.

The Planets and Their Satellites

Mercury	0	Saturn	18
Venus	0	Uranus	15
Earth	1	Neptune	8
Mars	2	Pluto	1
Jupiter	16		

Voyagers 1 and 2 changed our ideas about the solar system much the same way the discoveries of Uranus and Neptune changed people's thinking back in the eighteenth and nineteenth centuries.

2. The Discovery of Neptune

Ancient sky watchers knew there were five planets—Mercury, Venus, Mars, Jupiter, and Saturn. They knew this because they could see them in the night sky just as you and I can, and they could plot their motions. Of course, they also knew about Earth, but the ancients considered it to be the center of the universe, and they did not include Earth among the planets. In the 1500s Polish astronomer Nicolaus Copernicus suggested that Earth is a planet and that all the planets travel around the Sun. However, that idea was not widely accepted until the late 1600s.

The most distant planets—Uranus, Neptune, and Pluto— are too dim to be seen with the unaided eye. They were not discovered until long after the telescope had been invented.

Uranus was discovered in 1781 by William Herschel, a German-born astronomer who lived most of his life in England. The discovery created quite a sensation. The planet was twice as far from us as Saturn. All at once the solar system as it was known at that time doubled in size. Before the discovery of Uranus people had thought Saturn marked the end of the solar system.

About sixty years after Uranus's discovery, people had to make a few more changes in their thinking. Neptune was discovered in 1846. It was even farther away than Uranus. (Pluto was not discovered until 1930.) Although a telescope was needed to see Neptune, the presence of the planet was suspected even before the planet was seen.

After astronomers had plotted the motions of Uranus, they knew how fast it moved and its direction. Therefore they could predict exactly where it should be at any given time in the future. Within twenty years of the planet's discovery, astronomers began to notice that Uranus was not moving quite the way calculations said it should. Something was affecting the motion of Uranus. Some people believed there must be a planet-sized object out beyond Uranus that

John Couch Adams

was causing variation in the motion of the planet. One person who believed this was John Couch Adams, a student at Cambridge University in England. Adams decided that as soon as he graduated, which would be in 1843, he would try to figure out where the missing object might be.

14

After two years of carefully studying the motions of Uranus, Adams believed he could prove there was a planet out beyond Uranus. Also, he had figured out where it must be located. In October 1845 Adams went to Sir George Airy, who was astronomer royal (the chief astronomer in England), and asked him to search the suspected part of the sky.

Airy was not convinced that Adams was correct. He thought there were probably other reasons for the disturbance in the motion of Uranus, although he never said what those reasons might be. Besides, Airy said, it would take years to find such a planet, and there were other more important chores to be done. For eight months nothing was done about searching for Adams's planet.

Meanwhile, Urbain J. J. Leverrier, a French mathematician, had also set out to find the solution to the Uranus problem. In June 1846 he published a paper stating where the "missing planet" should be found. Adams read the paper and saw that he and Leverrier agreed about the location. Once again Adams asked the people at the observatory in England to look for the planet, which they did. For some reason, however, the astronomers did not keep careful records of their observations, and although they must have seen the planet, they did not know it. So no announcement was made.

At the same time, Leverrier got in touch with Johann G. Galle at the Berlin Observatory in Germany, asking him to

Urbain J. J. Leverrier

Johann G. Galle

turn his telescope to a certain part of the sky and to search for the planet. Galle received the letter September 23, 1846. That night he found Neptune.

The French thought the new planet should be called Leverrier. The English did not agree. After much arguing, both the French and English decided they would call it Neptune, after the Roman god of the sea.

It's hard to say who should get credit for the discovery of Neptune. You could say Galle made the discovery, since he was the first person to see the object and to recognize it. You could also say Adams discovered it, or Leverrier, for both of them figured out the planet's location. Usually when discoverers of planets are listed, both Adams and Leverrier are given credit. Although Galle was the first to see the planet, he is usually not mentioned.

3. Neptune—The Planet

Until August 1989 very little was known about Neptune. We knew where it was and how fast it moved around the Sun. We also knew that it had at least two satellites. But we knew little else about the planet. From Earth, even through the best telescopes, it appears as a tiny, dim point of light. The planet is almost 3 billion miles from the Sun; Earth's distance is only 93 million miles. Neptune is so far away that it receives only one thousandth of the amount of sunlight that we get. Planet eight is a cold world; temperatures there hover around 330° below zero F.

Neptune is the fourth-largest planet in the solar system. It is much smaller than giant Jupiter and Saturn and a little smaller than Uranus. These are the diameters of the four large planets:

Planet	Diameter (in miles)
Jupiter	88,846
Saturn	74,898
Uranus	31,763
Neptune	30,775

While Neptune is the smallest among the giant planets, it is a lot bigger than we are. Earth's diameter is 7,927 miles—almost four Earths could be placed side by side into the diameter of Neptune. The large planets are called gas giants because they are made mostly of gases.

Earth is quite different from the gas giants. A large part of Earth is rock. It also has a central core of nickel and iron. The large amounts of rock, nickel, and iron make Earth very dense and solid. Earth is said to have a density of 5.52. Water has a density of 1. A density of 5.52 means our planet weighs 5.52 times the weight of an equal volume of water. The density of Neptune is 1.64. A density of 1.64 means the planet must be made of materials whose average density is just a bit greater than that of water. The other gas giants have densities even lower than Neptune's. For the four giant

EARTH

From left to right: Jupiter, Saturn, Uranus, and Neptune. Neptune is the smallest of the four gas giants. Its diameter is almost four times the diameter of Earth.

planets to have such low densities, they must be made mostly of gases and have only small amounts of dense rocks and metals.

The most abundant gases in Neptune's atmosphere are hydrogen and helium. Smaller amounts of methane in the upper reaches of the atmosphere give Neptune its blue-green

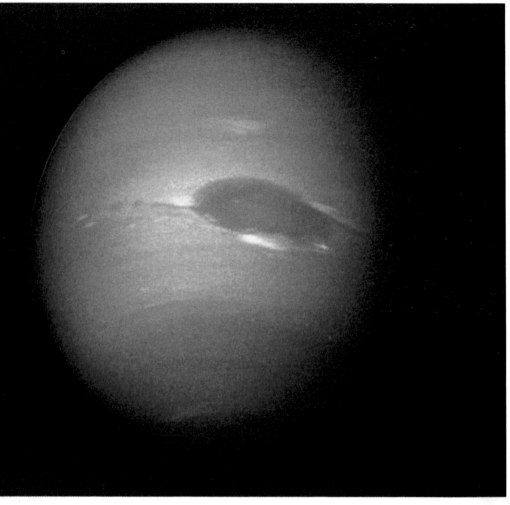

This color-enhanced picture shows Neptune's clouds in different colors. The highest are white and pink, the blue are a bit lower, and the lowest clouds are green. The dark-blue color of the Great Dark Spot tells us that the formation is deep in the atmosphere.

color. When sunlight hits the atmosphere, most of the colors in the sunlight are absorbed by the methane. The blue of sunlight is reflected to us.

When Voyager moved closer to the planet, it saw wispy white clouds of methane ice crystals floating above the lower layers of the atmosphere. Scientists knew the clouds were high because they cast shadows on the lower layers.

Streaks of high clouds catch sunlight (coming from the left) and cast shadows on the cloud deck below. The clouds are thirty to forty miles above the main cloud layer.

Voyager also saw a violent dark storm churning just south of Neptune's equator. The storm looks similar to a hurricane on Earth, but it is much bigger. In fact the storm is as large as the whole planet Earth. JPL scientists named it the Great Dark Spot. The spot seems to move westward at more than

The large and small dark spots show clearly here. The fast-moving white cloud patch called the Scooter can be seen between them.

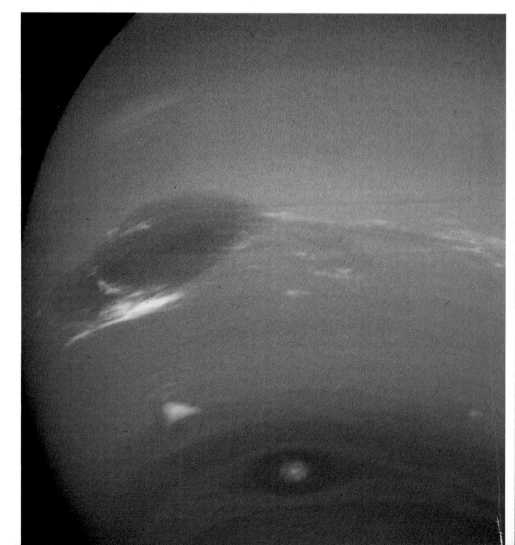

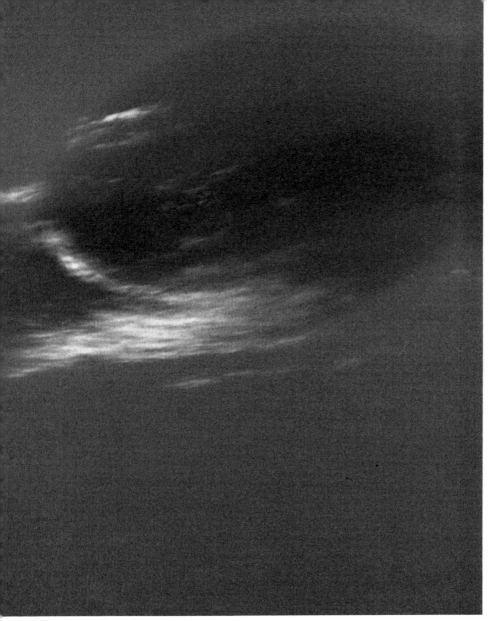

In this closeup of the Great Dark Spot we see white clouds of methane ice crystals above the edge. The spot is a violent storm system that rotates counterclockwise. It changes in size with the passage of time.

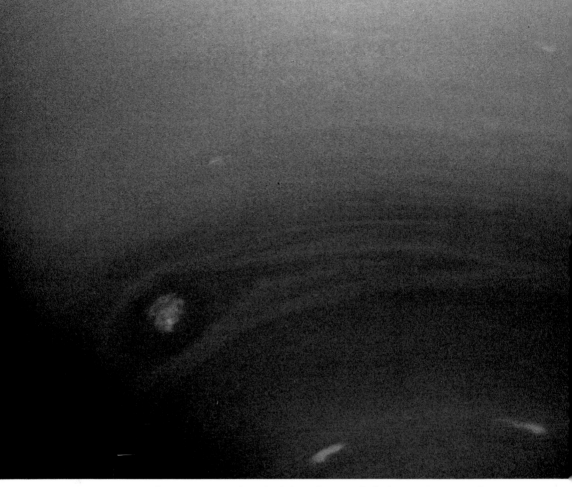

In this view of the southern hemisphere, the small dark spot is capped by white clouds. Farther south are streaks of white clouds.

1,300 miles an hour. Below the Great Dark Spot is another, smaller dark spot. The smaller spot seems to move about 200 miles an hour in the opposite direction. The cause of these great storms and their various motions remains a mystery. Between the two spots is a small white patch, or cloud. It was named the Scooter because it moved so much faster than the other clouds.

Scientists could see changes in the planet's atmosphere over a period of several hours. They watched clouds grow and diminish, dark streaks appear and then fade out. Neptune is certainly a planet with an active weather system.

As Voyager got close to Neptune, it passed through a field of electrified particles. They caused static interference in Voyager's radio signals. JPL scientists knew that the particles must have been trapped in a magnetic field surrounding the planet.

Electrified particles—electrically charged protons and electrons—stream out constantly from the Sun. The particles move at high speed in all directions. They stream past Earth and past the other planets. Many of the particles become trapped and held in the magnetic fields of the planets. The stream of high-speed particles is called the solar wind.

Voyager proved that the solar wind is still quite strong even as far away as Neptune. It also proved that the magnetic field of Neptune is strong enough to trap some of the solar wind particles.

Before Voyager, scientists doubted that Neptune had a magnetic field. If it did, they thought it would probably be very weak. Voyager found that the magnetic field of Neptune is similar to ours. It is weaker than the fields of Jupiter, Saturn, and Uranus.

Voyager's discovery of Neptune's magnetic field helped us learn more about the planet's interior. The presence of a magnetic field implies that a planet is made of layers of materials. One of these layers will be molten or semisolid. And one will be solid. As the planet rotates, the semisolid layer slides over the solid one. The movement of one layer over another produces an electric current. This internal electric current generates a magnetic field that surrounds the planet.

Neptune does not have a surface in the sense that Earth does. On Earth there is a sharp division between the gaseous atmosphere and the solid surface. (Oceans cover seventy percent of our surface, but they are only a few miles deep at most, while the solid crust is sixty miles deep or more in some places.) On Neptune, there probably is no sharp dividing line. Many scientists now think that Neptune's atmosphere becomes thicker and hotter toward the center of the planet. Gravity increases as you move closer to a planet's center. It pulls on the molecules of atmospheric gases and packs them together. Partway toward the center of Neptune, there may be a thick, gooey layer made of a mixture of atmospheric gases, water, and melted rock. At the center there may be a small solid core of rock and metals.

Scientists also used Neptune's magnetic field to measure accurately the rotation of the planet. The magnetism of a planet is strongest at its magnetic poles. Whenever Neptune's north magnetic pole turned toward Voyager, a strong magnetic pulse was recorded. The interval between two pulses—exactly 16 hours, 6.6 minutes—is the rotation period of the planet. While it rotates, Neptune also travels around the Sun, taking 164.8 years to complete one circuit.

4. The Satellites of Neptune

Before the Voyager flyby we knew about two of Neptune's satellites. Nereid, the smaller one, was discovered in 1949. We believed its diameter to be 180 miles. Now we know it is closer to 211 miles. In Roman mythology Nereid was one of the fifty daughters of Nereus, a sea nymph. Triton, the largest satellite, was discovered in 1846, the same year the planet was found. The satellite's diameter was believed to be 2,170 miles. It moves very fast, taking less than six days to go around Neptune. In mythology, Triton was a son of Poseidon, the Greek name for Neptune.

During the flyby Voyager 2 discovered six more satellites going around Neptune. The new satellites have been named

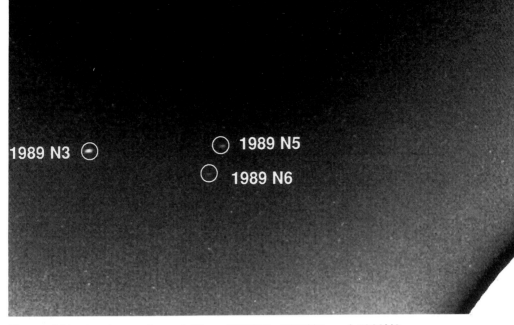

Three of Neptune's smaller satellites, 1989N3, 1989N5, and 1989N6, show up as faint smudges in this picture. Their images are blurred because their oribital speed is so fast—more than 25,000 miles an hour.

1989N1, 1989N2, 1989N3, 1989N4, 1989N5, and 1989N6, at least for the time being. Later they may be given more imaginative names. (Suggested names are on page 50.) All of them are small, ranging from 34 miles across (N6) to 260 miles (N1). All of them seem to be very black. JPL scientists believe we are seeing surfaces covered with frozen methane mixed with dark surface dust, probably carbon.

The probe also took a closer look at Triton. We now

1989N1 was the first of Neptune's satellites to be discovered by Voyager. Neptune's second-largest satellite, it measures about 260 miles across. It is lumpy—irregular in shape. Though the photo is very grainy, groovelike structures can be seen in the bright areas, and a large crater (about half the diameter of the satellite itself) appears on the lower right.

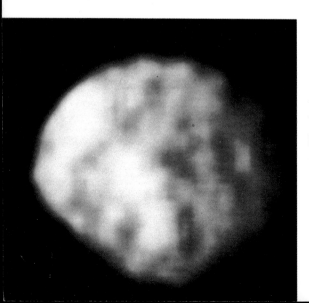

1989N2 is not exactly round; it is 130 miles by 118 miles in diameter. The surface appears to be pocked by large craters twenty to thirty miles across.

estimate it has a diameter of 1,678 miles, much smaller than previously believed—about three quarters the size of our Moon. Voyager's instruments revealed that the temperature of Triton is 390° below zero F. That makes Triton the coldest place ever observed in the solar system. (The coldest anything can be—the absence of all heat—is 459° below zero F., which is called absolute zero; so Triton is cold indeed.) At

This view of the southern hemisphere of Triton shows a variety of features: The lighter, pinkish-tan colored area is the south polar cap. North of the polar cap is an area of shallow, circular depressions and ridges. Long ridges mark out a gigantic "X" and "Y."

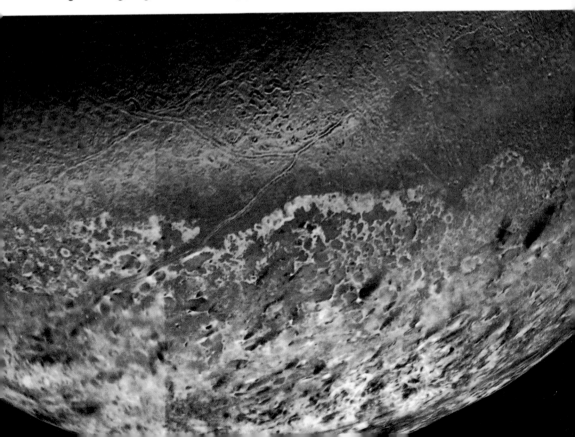

the time of Voyager's visit, Triton was some 2.8 billion miles from the Sun. In part, this extreme distance accounts for the low temperature. Also, much of the sunlight that does reach Triton is reflected by its ice-covered surface. (The hottest part of the solar system is Venus, which has a temperature of 860° F.)

Voyager's clear pictures of Triton reveal a satellite that is unlike any of the other sixty satellites in the entire solar system. Its surface has been compared to that of a cantaloupe, full of shallow depressions and ridges. It is relatively free of clearly defined large craters. The lack of craters has led many to believe that the features we see were formed fairly recently. That is because most cratering of satellites occurred some four billion years ago during a time of intense meteorite bombardment. More recent activity such as wind erosion has filled in craters and worn them down, so the surface has been smoothed out somewhat. Some of the features of Triton may have been formed by a kind of volcanic activity that has never been seen before.

Many observers think that ice volcanoes have formed there and may still be erupting from time to time. Triton is so cold that nitrogen gas is frozen out of the satellite's

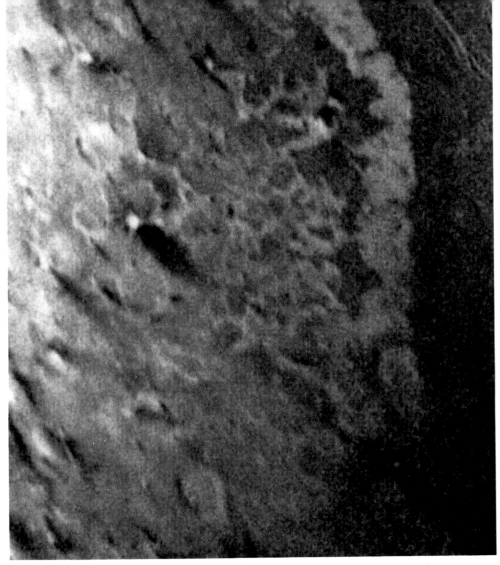

This false-color closeup of the south polar cap of Triton shows a huge ridge to the right. The polar frost is probably made of frozen nitrogen and methane. The black streaks spotting the frost may be nitrogen ice mixed with interior dark matter that was blown out in geysers by small ice volcanoes. Or they may be dark particles stirred up by whirling dust devils.

This smooth, round basin on Triton's surface is 120 miles across. It may have been formed by a huge ancient ice volcano that erupted and spread subsurface material (frozen methane, ammonia, and water ice) over the area, forming a frozen lake. The wall around the basin is about 600 feet high.

The crater near the center of the "lake" was probably formed when a large chunk of cosmic material (possibly an asteroid) crashed into Triton.

thin atmosphere. It falls as snow onto the surface. There may also be frozen nitrogen and perhaps methane several feet below the surface. Slight heating, perhaps caused by internal pressure or by some unknown heat source, may change the solids into liquids. The pressurized liquids explode through small openings in the satellite's crust. Geysers of the liquefied gases are thrown upward at speeds of one hundred miles an hour, carrying fine dark particles, mostly

A computer made this view of the basin shown in the previous picture. The features are exaggerated, but they give a good impression of what the flat basin would look like if you were standing in it. You can see the wall around it, and the impact crater toward the left.

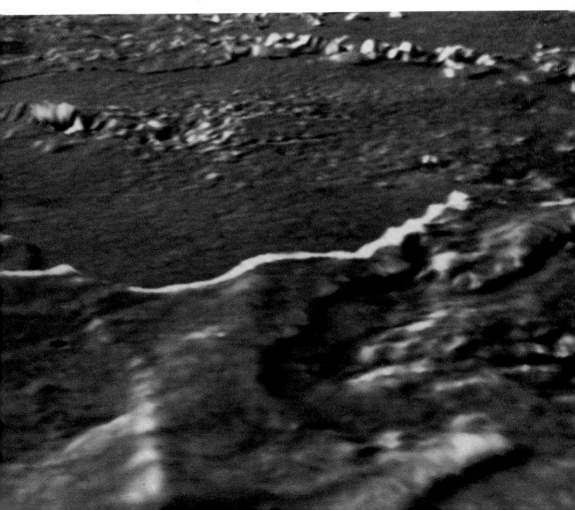

This photograph shows what scientists believe might be the plume of an active ice volcano on Triton, or perhaps the funnel of a swirling dust devil. The tall dark plume (between arrowheads) can be seen drifting downwind toward the right.

carbon, with them. The particles are carried along forty or fifty miles by the satellite's surface winds before they freeze solid and fall back to the surface, creating the long dark streaks we see. That is the most popular theory. Others have suggested that the dark plumes might be caused by "dust devils"—small intense tornadoes that lift columns of carbon particles off the surface. The particles are then blown downwind and settle to the surface in streaks. Careful study of the photographs indicates that the geysers or dust devils are still active.

Triton is full of fascinating mysteries. Prominent among them is the origin of the satellite. Nearly all the satellites of the planets, and all of Neptune's other satellites, move around their planets in the same direction that the planets rotate. Triton is different; it goes in the opposite direction. Also, its orbit is tilted 23.2° in relation to the equator of Neptune, while the orbits of most of the other satellites are closer to the equator. For these two reasons many believe that Triton did not form out of the same gas cloud that gave birth to Neptune. At one time Triton may have been a free-flying object—perhaps an asteroid—in its own orbit around the Sun. During a close approach to the planet, the satellite may have been captured by Neptune's gravitational field.

Nereid is about 3.5 million miles from Neptune. It is the planet's most distant satellite. Its orbit is tilted 27.6° to Neptune's equator. Because of this, and also because of Nereid's great distance, scientists think that it too is a captured object. Once thought to be the planet's second-largest satellite, we now know it is the third-largest. 1989N1 is larger; its dark surface prevented astronomers from discovering it earlier. When Voyager made its close flyby of Neptune and Triton, it was too far away to get good pictures of Nereid. However, scientists suppose that the surface of Nereid is similar in many ways to that of Triton.

Voyager did not get close enough to get a very good picture of Nereid, Neptune's third-largest satellite. But even from this blurry picture, scientists were able to get a more accurate measure of its size—about 211 miles across.

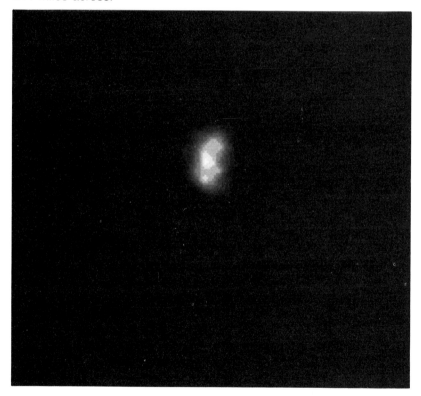

5. The Rings of Neptune

Long before 1989, astronomers suspected that Neptune might have rings. In fact, twenty-one years earlier, in 1968, Edward Guinan, an astronomer at Villanova University in Pennsylvania, saw their effect, although he could not see them. He was part of a research team in New Zealand to study the atmosphere of Neptune. New Zealand was the best place for viewing Neptune at that time. Guinan knew the exact time that the planet would pass in front of a particular star. As it did, he would watch and record the changes in the brightness of the star as its light shone through Neptune's atmosphere. The changes would make it possible to get information about the depth and density of Neptune's atmosphere. However, before the outer edge of the planet reached the star, the star's light dimmed briefly. It was not clear what had caused the dimming. The changes in brightness were recorded on the kind of punch cards used in the computers of that time. During the return boat trip from New Zealand the punch cards became wet. This warped them so that they could not be put through the computer.

The cards were stored and forgotten until 1977, when rings were discovered around Uranus. Astronomers discovered those rings while watching Uranus pass in front of a star, as Guinan had done with Neptune. This reminded Guinan of his punch cards and the brightness changes he had seen nine years earlier. He and his students undertook to copy each of the cards—a long and tedious job. When these new cards were put through the computer, a partial cutoff of starlight was seen above the atmosphere, about 30,000 miles from the surface of Neptune. As the planet moved away, a similar dimming occurred on the other side. The observations implied that there was some kind of formation around the planet. Perhaps it was a ring.

And it was. In 1989 Voyager found not one ring, but four of them. Two are distinct, one is fuzzy, and one is very faint and hard to detect. In the two sharper rings there are chunks six to twelve miles in diameter. These provide the gravitation needed to hold together the small particles and keep the rings from disintegrating.

The main outer ring (1989N1R) is about 39,000 miles from the center of Neptune. This is the sharpest ring and the one that contains the largest particles. The inner bright ring (1989N2R) is just inside the main outer ring. It is also sharp.

Neptune has four rings. Two of the rings show clearly; the other two are made of very small particles widely separated and are difficult to see. (To get this photograph, the planet itself had to be overexposed. It is blocked out in the picture.)

It lies 33,000 miles from the center of the planet. Twenty-six thousand miles from Neptune's center is the inside diffuse ring (1989N3R). Its particles are small, and the ring may be wide enough to extend all the way to Neptune's atmosphere. Between the main outer ring and the inner bright ring is a sheet of very fine particles called the plateau (1989N4R).

The main outer ring appears to have sausagelike clumps in it—segments that are brighter and thicker than the rest of the ring. Rings are made of billions of separate particles that are affected by the motions of the satellites. Ring particles also move, shift about, and reshape themselves as they collide with larger particles and sometimes even with small satellites. The sausages of the main outer ring are probably temporary. In later years if other space probes look at the ring system, the sausages will probably not be there.

Of the nine planets, only the four large gaseous planets have ring systems: Jupiter has three, Saturn has thousands (usually grouped into seven main rings), Uranus nine, and Neptune four.

Many scientists believe that all the planets formed from a great gaseous cloud—the same cloud of gas and dust that gave birth to the Sun. The particles of gas and dust in the

Neptune's outer main ring (1989N1R) contains bright arcs, or sausagelike clumps, where the ring particles are larger and more densely packed. These clumps move and change with the passage of time.

cloud were drawn together by gravity. Most of the material eventually clumped together into a huge mass that became the Sun. Smaller, leftover amounts became the planets and most of the satellites of the planets. There is still a lot of dust and gas in interplanetary space. Close to the planets the density of the material increases. In the vicinity of the largest planets, huge numbers of these particles may have been held together and shaped into rings by the strong gravitation of the planets. "Shepherd satellites," which sometimes lie in orbits close to the rings, may also help, through their own gravity, to shape the rings and hold them together.

Very likely, the gravitation of the planet itself disrupts these rings. It may force the particles to clump together, or it may pull them toward the planet. That is what may be happening with the inside, diffuse ring of Neptune.

Another theory holds that rings may be the remnants of satellites' breakups. Eventually (after billions of years), for example, our Moon may move in closer to Earth. The strong gravitation between the two bodies would then pull the Moon apart. It would shatter into small particles, which would become shaped into a ring encircling the planet.

6. Voyager Heads for the Stars

Voyager 2 and Voyager 1 have given us a library of information about the major planets—more than we have had time to understand completely. Voyager 2 is now on a path that will take it out of the solar system, as is Voyager 1. Many of the instruments aboard Voyager 2, as well as its nuclear electric generator, are expected to last well into the twenty-first century. All that time it will be gathering information about space itself and sending it back to Earth. Perhaps the probe will find where the solar system really ends—the region where the effects of the Sun, such as the solar wind, can no longer be detected.

Beyond lie the stars. Also, there may be other planetary systems out there. If so, creatures on faraway planets may someday recover the Voyagers. When they do, they'll find gold records containing sights and sounds of Earth. Should they play the records, they will find 118 pictures that can be developed from pulses, produced as the record spins, showing people, cars, and buildings. They will hear the sounds of 54 different languages, operas, and country music, as well as the sounds of machinery and traffic. Let's hope the Voyagers are found by intelligent extraterrestrials and that they play the records. Let's hope the beings out there will enjoy the sights and sounds of planet Earth.

Voyager 2 took a backward look at Neptune and Triton as the probe began its endless journey out beyond the solar system.

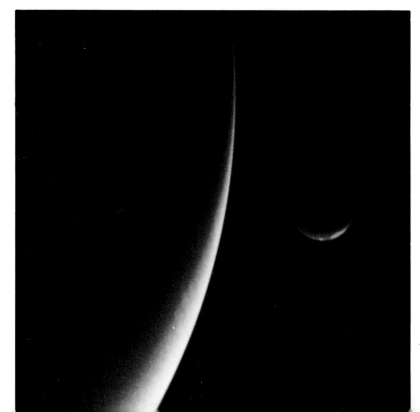

Appendices

Neptune and Earth

	Neptune	*Earth*
Rotation	16 hours 6.6 minutes	23 hours 56 minutes
Revolution	60,190 days (164.8 years)	365.24 days
Velocity in orbit	3.37 miles per second	18.51 miles per second
Inclination of equator to orbit	28°8	23°5
Mean Distance from Sun (in millions of miles)	2,799	92.96
Volume (Earth = 1)	57.7	1.0
Mass (Earth = 1)	17.15	1.0
Density (Water = 1)	1.64	5.52
Diameter (in miles)	30,775	7,927
Atmosphere (main parts)	Hydrogen, Helium, Methane, Ammonia	Nitrogen, Oxygen
Satellites	8	1
Rings	4	0

Neptune's Satellites

Name	Suggested Name*	Discoverer	Year	Distance from Neptune (in miles)	Diameter (in miles)
1989N6	Naiad	Voyager 2	1989	29,830	34
1989N5	Thalassa	Voyager 2	1989	31,070	50
1989N3	Despoina	Voyager 2	1989	32,620	93
1989N4	Galatea	Voyager 2	1989	38,530	99
1989N2	Larissa	Voyager 2	1989	45,730	118
1989N1	Proteus	Voyager 2	1989	73,100	260
Triton		W. Lassell	1846	220,440	1,678
Nereid		G. Kuiper	1949	3,423,200	211

*These names are being considered by a special committee of the International Astro-nomical Union. If the committee approves them, they will become official.

Further Reading

Abelson, Philip H. "The Human-Voyager Configuration." *Science*, September 15, 1989, p. 1161.

Beatty, J. Kelly. "Getting to Know Neptune." *Sky & Telescope*, February 1990, pp. 146–155.

Beatty, J. Kelly, and Andrew Chaikin, eds. *The New Solar System*, 3rd ed. Cambridge: Sky Publ. Corp., 1990.

Branley, Franklyn M. *Saturn: The Spectacular Planet*. New York: Thomas Y. Crowell, 1983.

———. *Uranus: The Seventh Planet*. New York: Thomas Y. Crowell, 1988.

———. *The Nine Planets*, rev. ed. New York: Thomas Y. Crowell, 1978.

———. *Jupiter: King of the Gods, Giant of the Planets*. New York: Lodestar Books, 1980.

Goldreich, P., et al. "Neptune's Story." *Science*, August 4, 1989, pp. 500–501.

Gore, Rick. "Neptune: Voyager's Last Picture Show." *National Geographic*, August 1990, pp. 34–47.

Kerr, Richard A. "Triton Steals Voyager's Last Show." *Science*, September 1, 1989, pp. 928–930.

Kinoshita, June. "Neptune." *Scientific American*, November 1989, pp. 83–91.

Miner, Ellis D. "Voyager 2's Encounter with the Gas Giants." *Physics Today*, July 1990, pp. 40–47.

"Voyager's Last Picture Show." *Sky & Telescope*, November 1989, pp. 463–470.

Index

Numbers in *italics* refer to illustrations.